Engineering

# Wings to the Sun

By Cherish Erb-White, Leah Saracco,
Alyssa Westrom and Kennedy Johnson

*Dedicated to those who develop and apply world-class science, technology, and engineering.*

Text by: Alyssa Westrom, Cherish Erb-White, Leah Saracco, Kennedy Johnson, and with Glenn McIntyre and Dr. Ellen Cavanaugh

Images by: Alyssa Westrom, Cherish Erb-White, Leah Saracco, Kennedy Johnson and taken from Creative Commons.

*ISBN: 978-0-359-42000-1*

Grow a Generation
Sewickley, PA 15143
www.growageneration.com

Any and all profits from the sale of this book benefits The American Ceramic Society.

They knew that the mythological Icarus couldn't have flown too close to the Sun. The Sun is 93 million miles away, and it's at least 50 miles to escape the gravity and atmosphere of Earth.

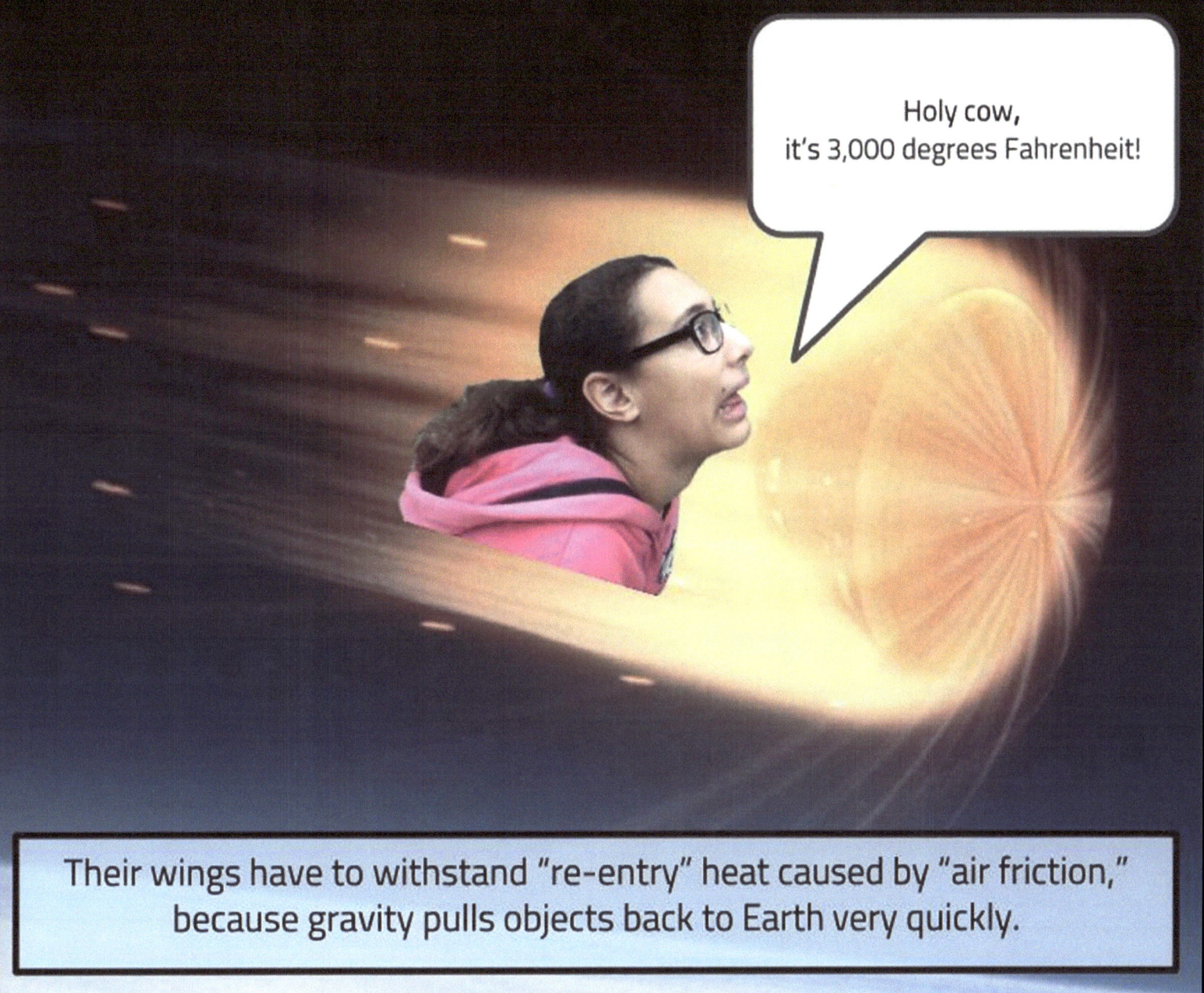

They looked for what other engineers had designed to survive the radiation beyond the atmosphere and the extreme heat of falling to earth.

They found NASA's Space Shuttle and the Parker Solar Probe both had to deal with leaving the atmosphere. The Space Shuttle had to re-enter the atmosphere and withstand the more than 3000°F heat caused by re-entry.

The Parker Solar Probe is flying multiple times around the Sun to study the star up close. Both the Space Shuttle and the Parker Solar Probe use ceramics to protect the equipment and, in the Space Shuttle's case, people, from heat and radiation.

Destiney, one of our school's Sun Superheroes, placed her name aboard the Parker Solar Probe before it left earth.

# Typical Refractory Elements

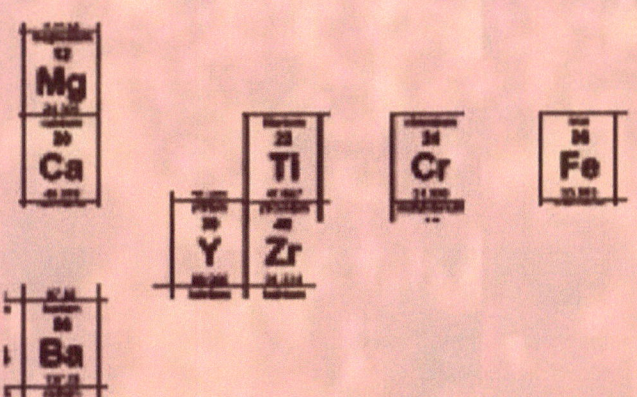

## What are ceramics?

Ceramics are great materials created from unique combinations of a few elements from the periodic table. They are shaped and then baked at high temperatures. The high temperatures cause a chemical reaction that makes the material incredibly strong and able to withstand extreme temperatures without breaking or melting.

Ceramics are fascinating! The clay begins as something you can shape and mold, and then when fired, it becomes hard. It comes in different varieties that require either low or high temperatures. Each ceramic recipe uses many different scientific properties.

Some people think ceramics are the best building material.

- Ceramics can withstand VERY high temperatures.
- Ceramics can be made to be electric insulators or semiconductors or super-conductors.
- Ceramics can be magnetic or non-magnetic.

Daedalus and Icarus's flight over the island Icaria ancient Roman-Greece pottery.

Have you ever held something made of ceramic? There are probably a lot of tiles, plates, and mugs in your home that are made of ceramic. Ceramics have been used to make useful pots, beautiful vases, and great works of art, and have often been the secret to uncovering ancient history.

The ceramic tiles on the Space Shuttle were individually made and fit into place. Each tile is a different thickness and a different density. Temperatures on the shuttle reach several thousand degrees. Protecting the shuttle and the crew from such heat is very important.

Mrs. Melanie Houston is a 5th grade teacher at Baden Academy who uses an actual Space Shuttle Tile to teach math and science in her project The Houston Solution!

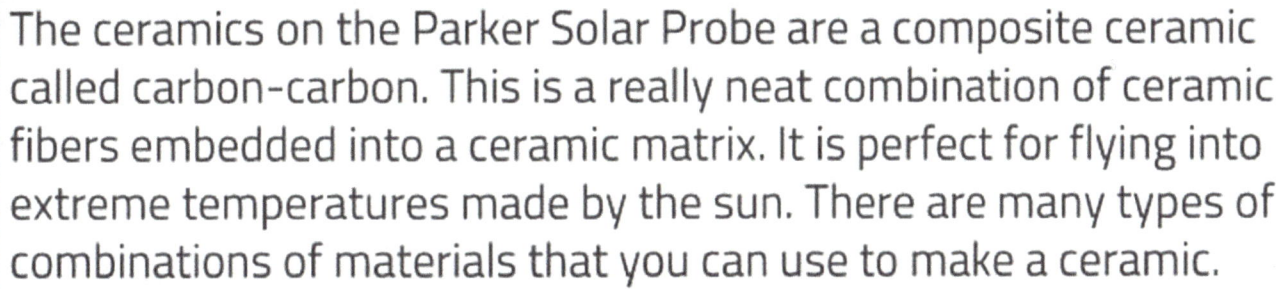

The ceramics on the Parker Solar Probe are a composite ceramic called carbon-carbon. This is a really neat combination of ceramic fibers embedded into a ceramic matrix. It is perfect for flying into extreme temperatures made by the sun. There are many types of combinations of materials that you can use to make a ceramic.

Destiney Campbell and Lizzie Hogue
Sun Superheroes

The engineers knew they wanted to make their wings out of ceramic, but they needed to decide what kind of wing design would be best. They studied wing designs made by brilliant engineers of history.

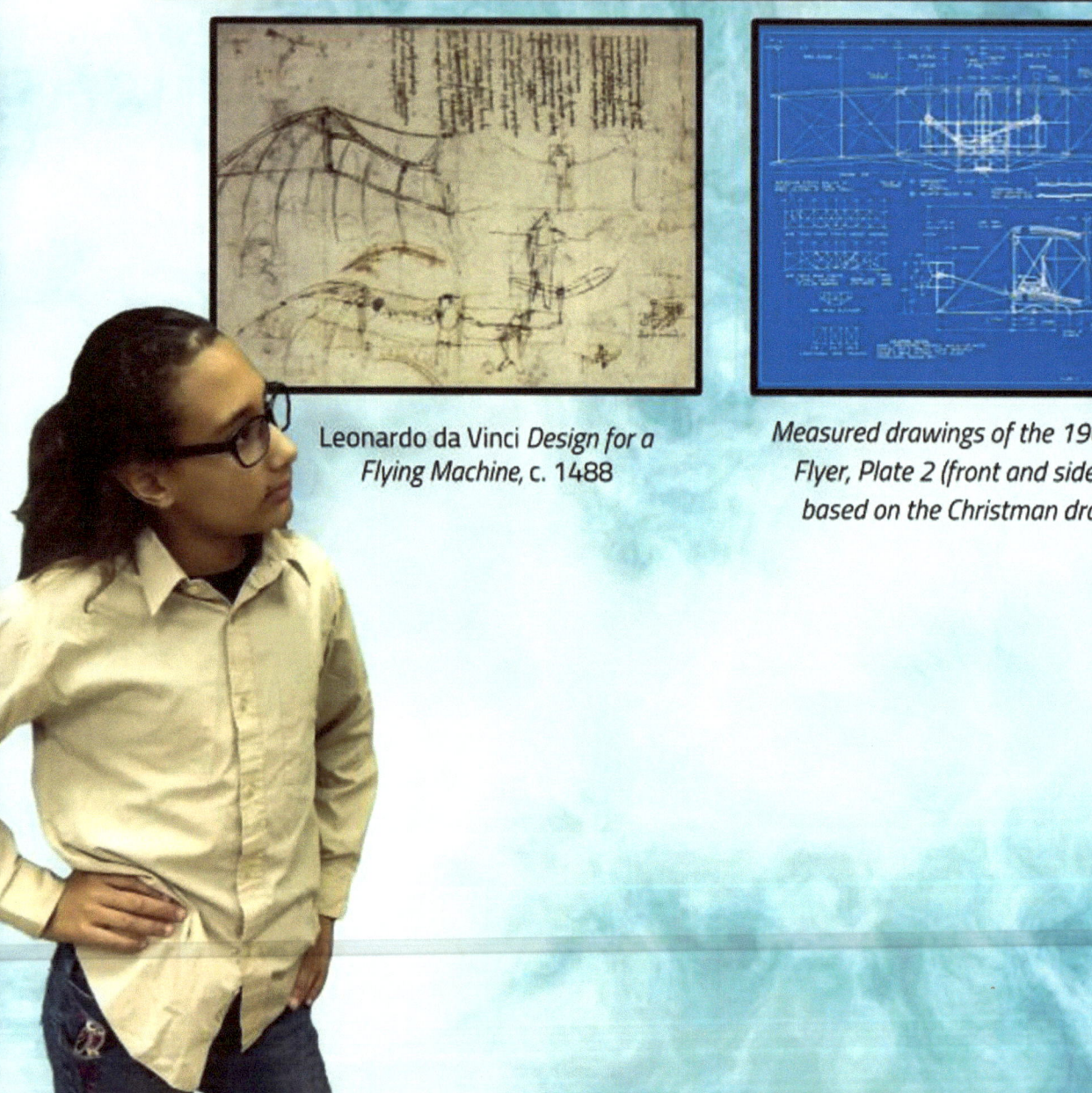

Leonardo da Vinci *Design for a Flying Machine*, c. 1488

*Measured drawings of the 1903 Wright Flyer, Plate 2 (front and side views), based on the Christman drawings.*

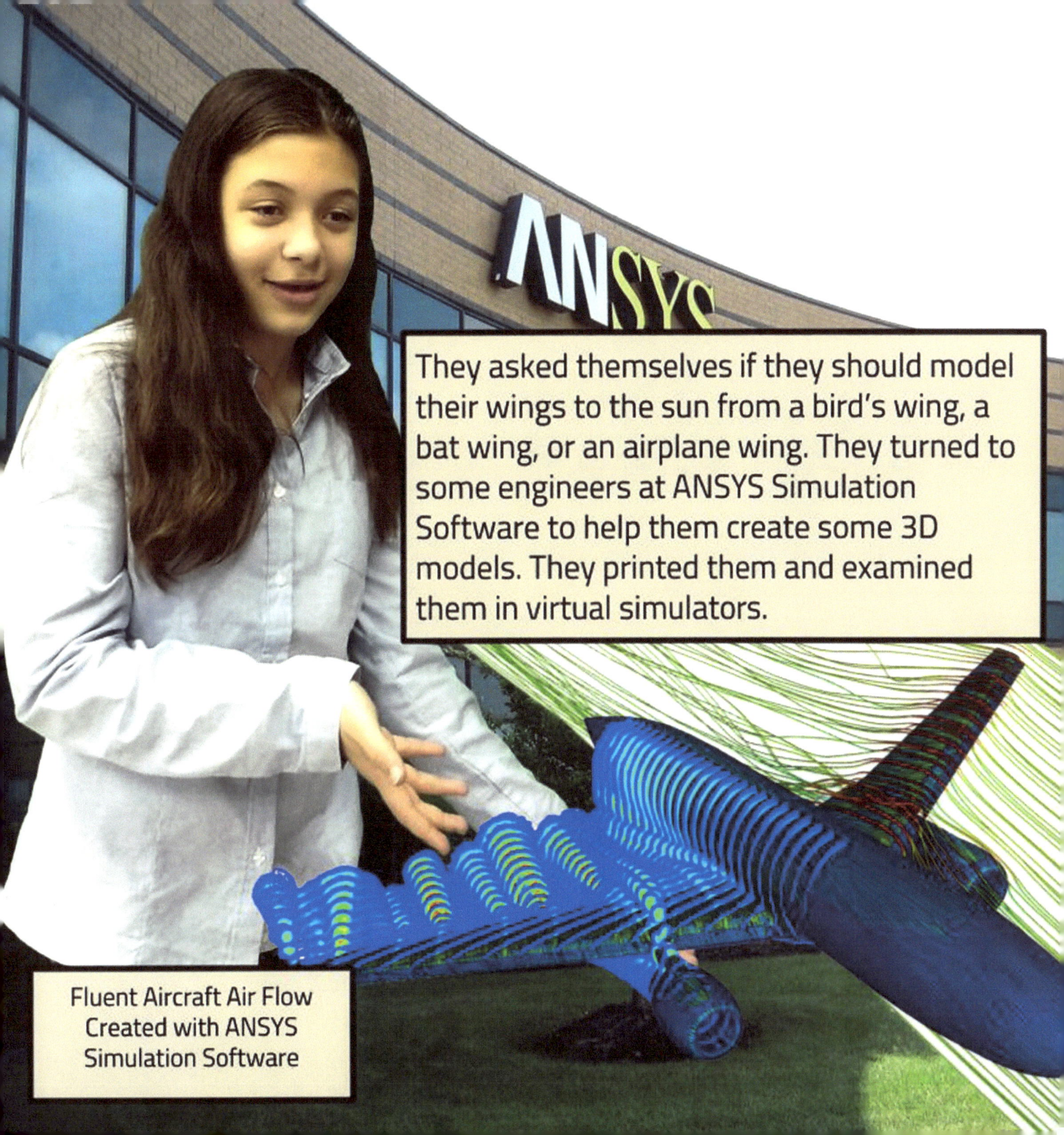

They asked themselves if they should model their wings to the sun from a bird's wing, a bat wing, or an airplane wing. They turned to some engineers at ANSYS Simulation Software to help them create some 3D models. They printed them and examined them in virtual simulators.

Fluent Aircraft Air Flow Created with ANSYS Simulation Software

The engineers began to identify some problems with their design ideas. They couldn't make the entire wing out of one solid piece of ceramic. It wouldn't be able to flex and move. Plus, they would need a really big oven to bake it.

The Engineers traveled to Swindell Dressler to learn about kilns (ovens to bake ceramics). They learned the fascinating technology behind heating the ceramics to the exact temperatures. They also discovered it is possible to build a kiln bigger than a football field where the cermics ride through on a type of tram car. That was impractical for just one set of wings!

A bird's wing is made of interconnected and layered feathers. They experimented with a design of multiple layers of feathers that could be flexible.

Individual ceramic feathers could be made to fit together on the wing. But would making those with the fine detail of feathers be hard? There is a huge difference between a brick and a feather! Each feather would have to individually cast and machined.

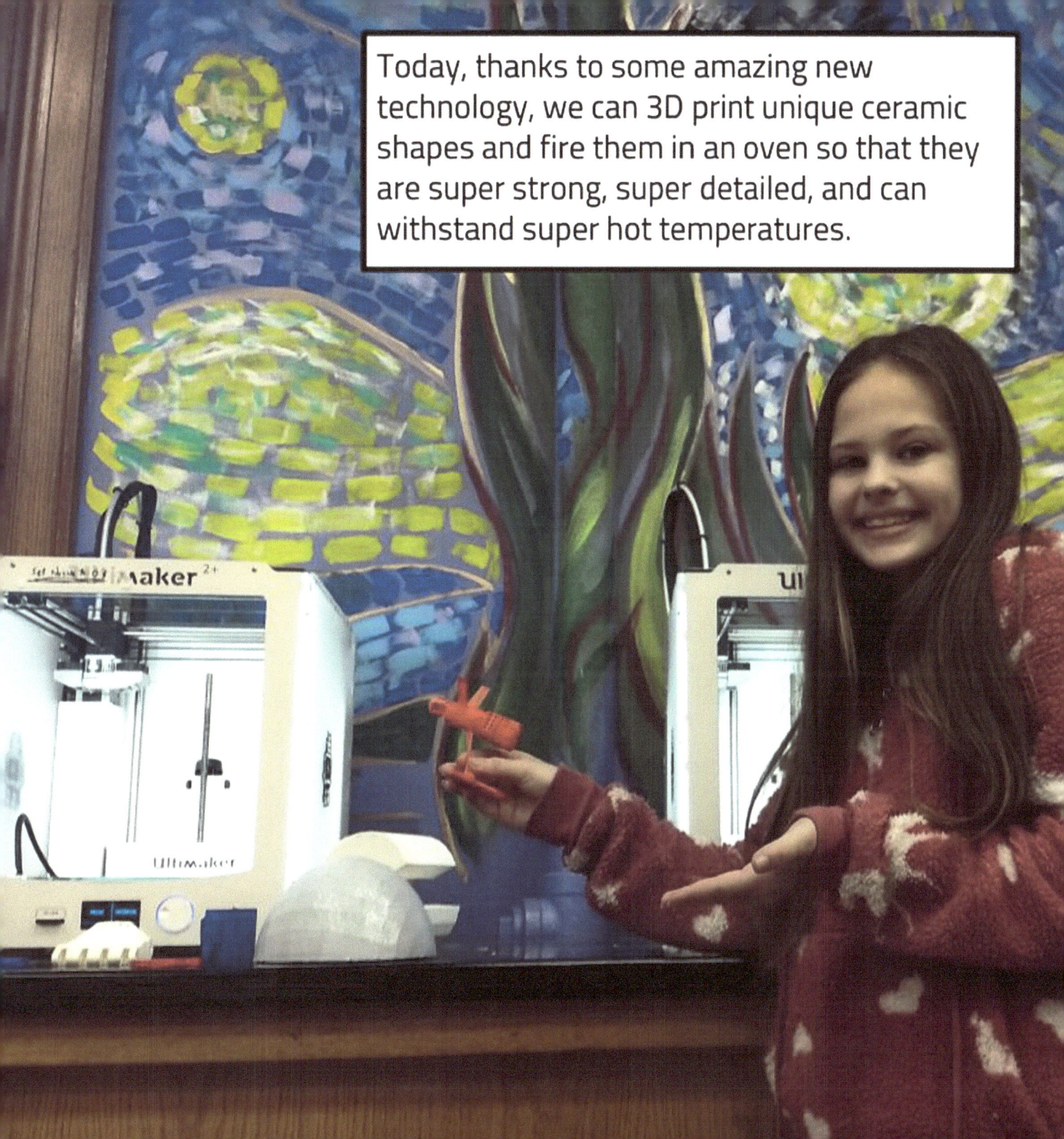

Most 3D printers are called "additive." That means they add one layer at a time on top of the previous layer to build.

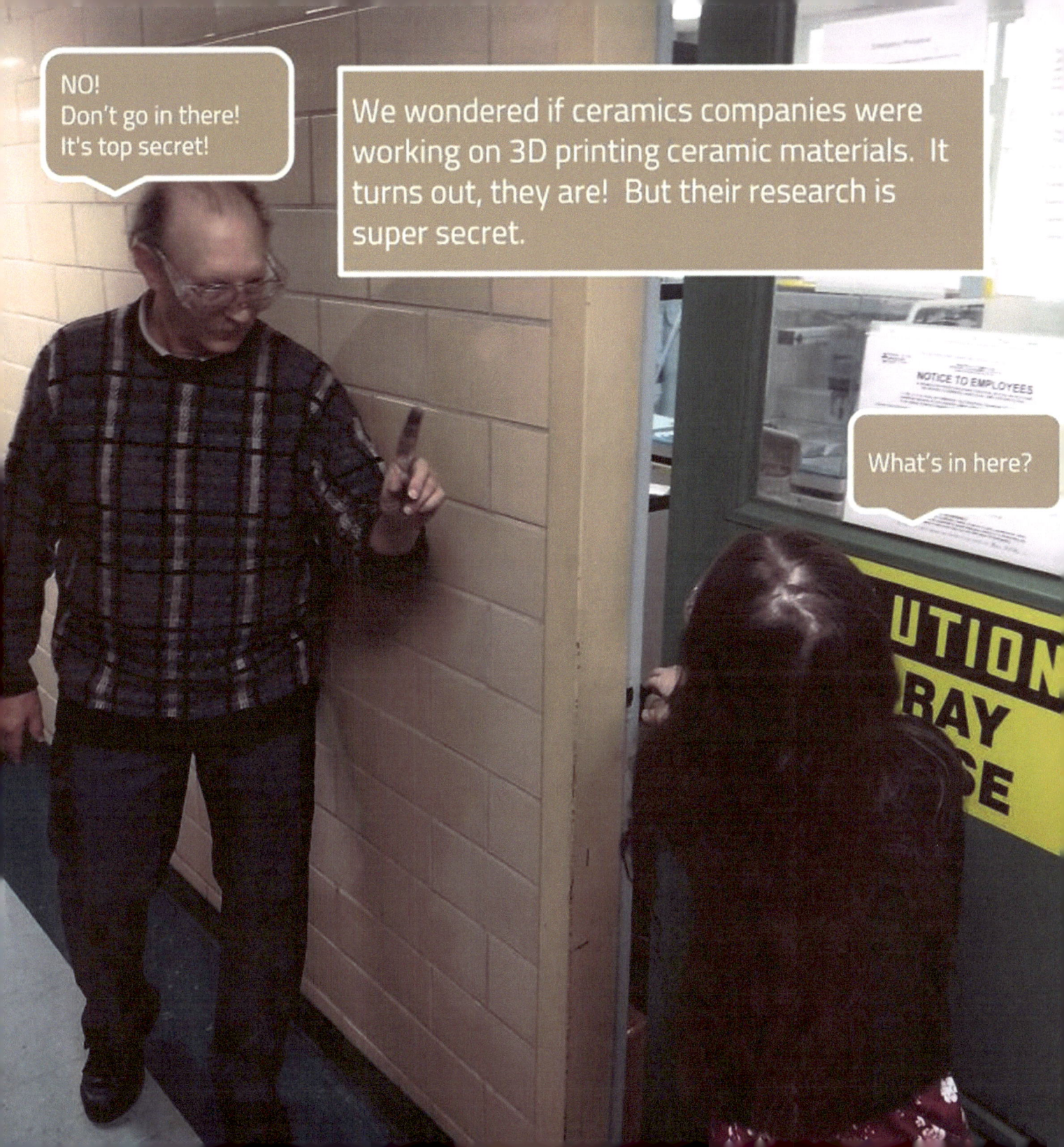

The engineers contacted ceramics engineers at Harbison Walker, the United States largest manufacturer of refractory products. (Refractory is a substance resistant to heat.)

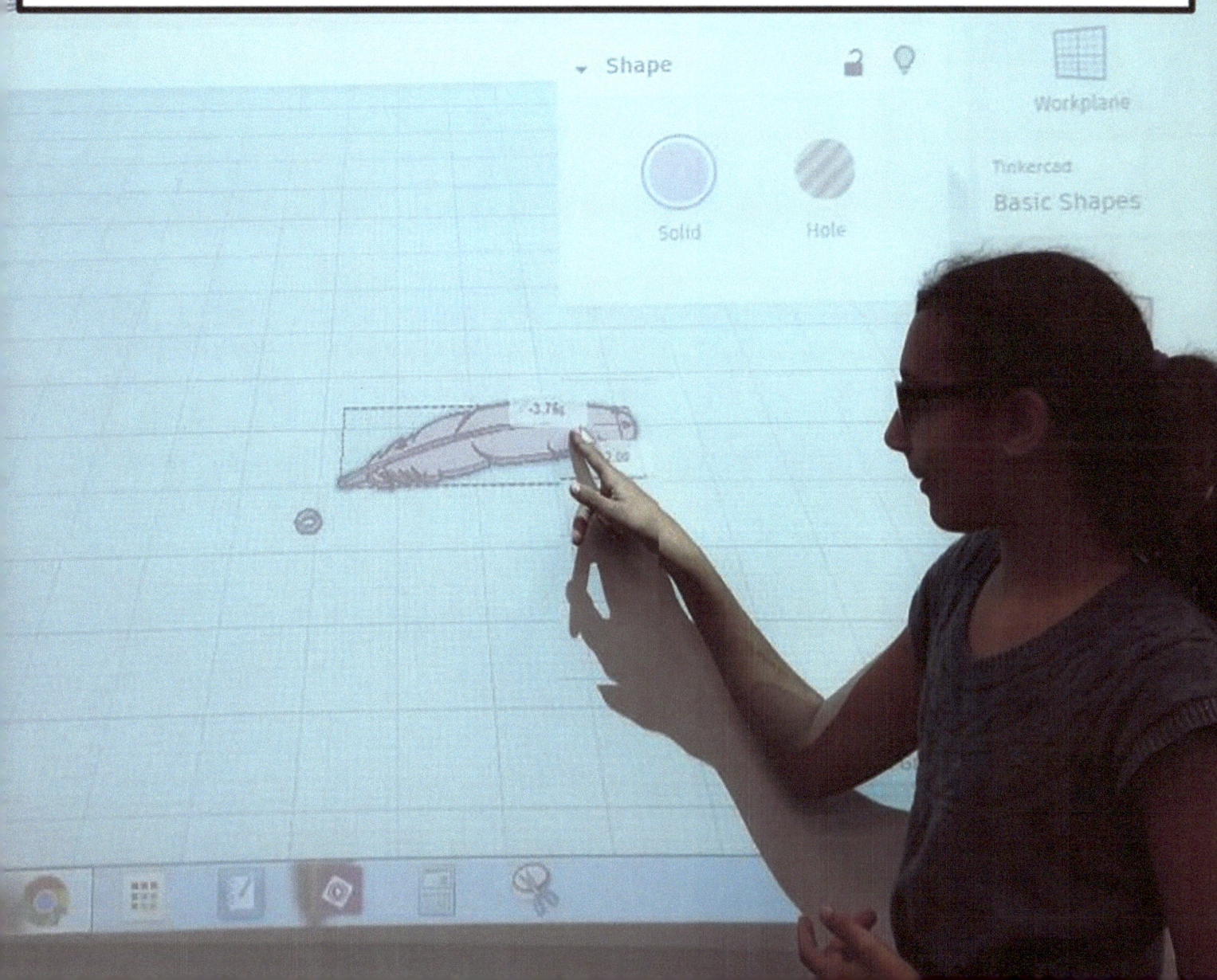

They were able to meet with Glenn McIntyre, one of the company's technology managers. He took them on a tour of some exciting advances happening in ceramics. The engineers showed Mr. McIntyre their design.

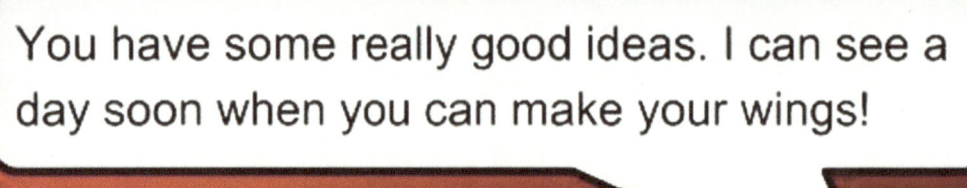

Glenn McIntyre is the Technology Manager of Bricks at the Advanced Technology and Research Center of Harbison Walker

Hi - We're Alyssa, Cherish, Kennedy, and Leah. We dream of the day we can design superhero wings and fly like Icarus, without falling!

The science of ceramics is real! The technology of 3D printing is real! The math to test in simulators is real! The possibility of engineering wings to the sun inspires us to learn more. We hope it inspires you too!

# Baden Academy Charter School

This public charter school in Western PA works to inspire personal excellence. They cultivate the inherent gifts and talents present in all children by providing a curriculum that integrates the arts and sciences in a highly interactive, hands-on environment.

# Grow a Generation

Grow a Generation partners with gifted and talented young people and teachers to make meaningful projects possible. Faculty, students, and student teams apply in their school to be accepted into the fellowship program. Once selected, they embark on a year-long odyssey to publish a book, create a digital artifact, or enter a STEM competition. Find out more at growageneration.com

www.ingramcontent.com/pod-product-compliance
Lightning Source LLC
Chambersburg PA
CBHW041301180526
45172CB00003B/924